THE
SEVERN BORE
AN ILLUSTRATED GUIDE

CHRIS WITTS

For my wife Susan

First published 2011

Amberley Publishing
Cirencester Road, Chalford,
Stroud, Gloucestershire, GL6 8PE

www.amberleybooks.com

Copyright © Chris Witts, 2011

The right of Chris Witts to be identified as the
Author of this work has been asserted in accordance
with the Copyrights, Designs and Patents Act 1988.

ISBN 978-1-84868-973-2

British Library Cataloguing in Publication Data.
A catalogue record for this book is available from
the British Library.

Typeset in 9.5pt on 12pt Celeste.
Typesetting by Amberley Publishing.
Printed in the UK.

Contents

The Severn Bore with May Hill in the distance.

The Severn Bore sweeps along the bank opposite Stonebench.

4

Introduction

A few years ago, BBC Radio 4 presented a series on the greatest natural phenomena of Britain. One of those featured in the series was the Severn Bore, and they asked me if I would talk about the bore as it passed Stonebench.

They wanted a night bore. A date was chosen and we arranged to meet at Stonebench. I groaned as I arrived at the chosen location, for it was drizzling and quite cold. Also, not a fan of night-time bores, I slowly became less enthusiastic about being there. Soon the female presenter arrived with her assistant and told me this was the last event in the series and how good the other phenomena had been.

We found a spot in amongst some bushes and waited. Both of them soon told me that they had never seen a bore before and were quite excited about this one. During this time of patiently waiting, I outlined what we may expect, thinking to myself that it could be a poor bore, in which case they would be disappointed.

As I indicated that I thought it was on its way, we stopped talking and absorbed the atmosphere around us. Everything became still and then, with the sound of an express train, the bore was suddenly upon us. Crashing through the trees with the sound of branches snapping, it raced past us with a roar. Even I was excited as I felt the hairs standing up on the back of my head.

The presenter was ecstatic, saying that this was the best natural phenomenon of the series. For me it was the same, and to this day I rate that bore as the best bore I have witnessed. For bore watchers, seeing a night bore is a relatively new experience, something that has caught on in latter years, but there is a secret as to how to get the best from experiencing one at night. (That secret will be explained later in the book.)

I didn't experience my first bore until I was a young married man with a family. This surprises most people, who know that I have worked on the River Severn since the age of sixteen. During those early years, the bore wasn't mentioned, simply the spring tides and a way of working the tides in the estuary and upriver from Gloucester to Worcester. I was working on estuarial tanker barges carrying petrol from Swansea inland as far as Worcester. Although we experienced the effect of the bore above Gloucester, it was just a surge of water without the wave.

For a Christmas present my brother gave me an old cine camera, and my first footage filmed was the Severn Bore. On a cold and dismal morning, the family and I went to Elmore Back to see the bore. I didn't know what to expect and didn't know much about how to use the cine camera, but I had a lot of beginner's luck. Shortly before the bore arrived, the stone barge *Riparian* – with well-known Severn expert the late Fred Rowbotham in the wheelhouse – appeared from upriver. I began to film and my luck was in, for I caught on film the *Riparian* riding the bore. I had that cine film put onto VHS tape, but having since moved house I now cannot find the tape, much to my annoyance!

The Severn Bore catches the imagination of people from around the world, with many emailing me for advice on where the best place is to see the bore. I have taken people from afar to see the bore and as it passes I look at their faces and see a slight disappointment. I ask them what did they expect and many times I get the reply, 'We expected to see a wave as high as a house.' I tell them that when that happens it would be time to run, because Gloucester would be underwater!

I talk to many people who live in Gloucester and I discover that a great many of them have never seen the bore. I guess they take it for granted and one day they will get around to finding the time to go and see it. If the much talked about barrage is ever built, this could signify the end of the Severn Bore as we know it today. That will be very sad, as our very own Severn Bore is one of Britain's greatest phenomena.

The Severn Bore

Why would crowds of people stand on the bank of a river in all weathers waiting, maybe for quite a while, before anything happens? The River Severn below Gloucester draws in hundreds of people from far and wide each time the media announces the forecast of a good bore.

They come in their cars, leaving them abandoned along roads close to the river, in many cases causing traffic jams and loss of tempers. Having found somewhere to park, they make their way to the river bank and join the others as they stand patiently waiting for the bore.

Many of them are first time bore watchers who don't really know what to expect. Indeed, many are ignorant of what the bore is. This becomes obvious by listening to their conversations, some saying it only occurs during springtime, some saying it only happens about six times a year.

There are a few false alarms, as their patience grows thin. Someone shouts out that they can see the bore in the distance, but usually it is their imagination playing tricks on them.

When the bore does finally arrive, there is no mistaking it. Usually it is of average height, not too large; thus many people leave muttering, 'Was that it?' They go back to their cars feeling disappointed, many never returning to see it again.

What we must remember is that watching the spectacle of the bore in the twenty-first century is a lot different than watching it a hundred years ago. Then the river had clear open views for miles, without the miles of foliage along the bank that can be seen today. From a painting by Charles Gere it can be seen that there was a lot of interest in the bore, with people on the banks cheering as men in two punts are waiting to ride the approaching bore. Today, we have the surfers, who are determined to ride the bore on their boards from Newnham on Severn to Gloucester. They come from all over the world to enjoy the experience – but not to everyone else's enjoyment!

Accompanying the surfers are several semi-rigid boats with their noisy outboard engines, riding backwards and forwards over the wave. They are there as safety boats for the surfers, but they do spoil the bore for the many hundreds of people watching it from the bank. Now they are being encouraged to stay back from the bore wave, thus allowing the bore watchers a chance to see the wave as nature intended.

Surfing the Severn bore is dangerous as many surfers get taken into the bank and through the overhanging trees and boulders that protect the bank from erosion. Some surfers are known to surf the bore at night. This is very unwise!

Many rivers around the world have this natural phenomenon called a bore. The River Severn is rated as the best for them in Europe, but the largest of them all is in China, on the Qiantang River. A visit to YouTube on the Internet offers many fine video clips showing the power of this bore. One aerial film actually shows people on the bank engulfed by the bore, with some washed into the river.

THE SEVERN BORE
by Charles Gere R.A. (1869-1957)
Purchased 1952

An oil painting of the Severn Bore by Charles Gere R.A. (1809–1957). The painting hangs in North Warehouse, the headquarters of Gloucester City Council.

An inflatable rides the bore as it passes the Lower Parting.

Chasing the bore with the safety boats.

Riding the bore at Weir Green.

A jet skier chases the bore at speed.

A boat of bore chasers churn up the Severn.

Cutting through choppy water, another boat follows in the wake of the Severn Bore.

The largest bore in the world on the Qiantang River in China.

The Qiantang Bore crashes over a bank.

What is a Bore?

In 1964, Fred Rowbotham penned a technical phrase to describe a bore: 'A bore is an undular surge wave formed at the foremost part of a tide that has run into a channel of particular shape and proportion, both in plan and in the shape of its bed.'

Bores are part of a tide, to be precise a spring tide, which is very large and appears in conjunction with a new or full moon. There are two types of tide, a spring tide and a neap tide (also known as a small tide). As an example, during a typical month, on the first day, the height of the tide at Sharpness could be 5 metres (16.5 feet) (a low neap tide) rising each day until the ninth day to a height of 9 metres (29.5 feet) (a spring tide). The tides then drop off down to 5 metres (16.5 feet) on the fifteenth day before rising to 10 metres (32.8 feet) on the twenty-third day. From this chart it can be seen that spring and neap tides occur twice a month, twice each day. Tides occur twice in a 25-hour period, each one moving forward on average by 25 minutes, with the average interval between high water of each tide being 12 hours, 25 minutes.

There is much misunderstanding about the term 'spring tide'. Many believe it to mean that the bore only appears during springtime. Nothing could be farther from the truth. Our spring tide has nothing to do with the season but is derived from the term 'to leap up high'. These high tides come in conjunction with the lunar cycle of the full and new moon. The new moon for seven months of the year will produce the highest tides, then for the next seven months it will be the full moon that produces the highest tides. This lunar cycle lasts for about nine years, which is why tidal predictions can be given for quite a few years in advance.

In the course of a year the highest tides occur during the spring (February to April) and autumn (August to October) equinox. This is when the media go crazy and predict a good bore, which brings people out in their hundreds to line the banks of the Severn, all hoping to see a good one. If only it was so simple! There are many factors, besides a high spring tide, which must be present before a good bore happens.

The dates and times of Severn bores are based on the Bristol Channel tide book for Sharpness. As a rule of thumb, a predicted tide height of 8 metres and above at Sharpness will produce a bore. Times shown in the book are always given as Greenwich Mean Time (GMT), thus during British Summer Time (BST) an hour must be added.

A careful study of the tide book will indicate that bores only occur between 7 a.m. and noon, and 7 p.m. and midnight. Another chart will have to be consulted for the location where you wish to see the bore to ascertain at what time the bore will arrive. Remember, the times are only predictions and very rarely does the bore arrive on time. It's better to arrive early than miss the bore by a few minutes. One thing is guaranteed: the bore will arrive. If you have been stood in the cold for a while without any sign of anything happening, just be patient, as the wait will be worth it.

The tide far out in the ocean is different to what we see at the seaside. As the tide approaches Europe from the Atlantic and is two hundred miles offshore, it comes into the shallower waters of the continental shelf. This has the effect of slowing down the tide but increasing its height.

The spring tide from the Atlantic enters the Bristol Channel, which is located between Pembroke in South Wales and Hartland Point in Devon. The distance between these two points is 50 miles, yet 85 miles further on, before meeting the constriction and rising riverbed of the Severn Estuary opposite Avonmouth, the width is only 5 miles. As the tide runs over the low sands, the sheet of water is headed by a small wave barely 50 millimetres (2 inches) high. If anyone is foolish enough to be out on the sands before the tide comes in, they are in imminent danger of being cut off. The sand banks are so low that the quickly rising tide soon covers them, pushing a small wave in front faster than you can get away from it.

Where to see the Bore

There are several factors required by nature to produce a good bore or indeed kill one. The term 'freshwater' is used quite often and refers to the amount of water flowing down the Severn above the normal summer levels. After a period of rainfall along the course of the river above Gloucester, the level of the river will rise: this is 'freshwater'. The amount of freshwater in the river determines the size of the bore and will help decide the best place to see this natural phenomenon.

Usually during winter months, the river is flowing higher than the summer level above Minsterworth, which will have the effect of reducing the size of the bore. Yet below Minsterworth this has the effect of increasing the bore and making good a spectacular wave. Places like Newnham on Severn, Arlingham, Framilode and Epney can be a good spot to view the bore during these conditions. With no freshwater in the river, the opposite is true; below Minsterworth the bore can look poor, yet above this location it can be excellent.

The added bonus of seeing the bore as far down the river as Newnham on Severn, Arlingham or Epney is that you will have time to drive further up the river and see it again, assuming that there are not too many people doing the same and causing parking problems. As the bore at times is only travelling at less than 10 mph, this will give you plenty of time to see it twice or even three times until it ceases at Gloucester.

It is becoming increasingly difficult to get a decent view of the bore from the banks of the Severn above Elmore Back. Many of the favourite locations are now lined with trees that obstruct the bore watcher from getting a clear view of the bore as it travels up the Severn towards Gloucester. Even at Stonebench, which has been a favourite for years, the bank is now lined with the quick-growing willow trees that are seen along most of the 220-mile course of the Severn.

The bore will look different depending on which part of the river you choose to go to see it. Below Sharpness the tide, as it comes into the estuary, appears with a succession of three or four long and very shallow unbroken waves with a smooth appearance. The Severn Estuary has the second highest rise and fall of tide in the world, the highest being (only just though) in the Bay of Fundy in

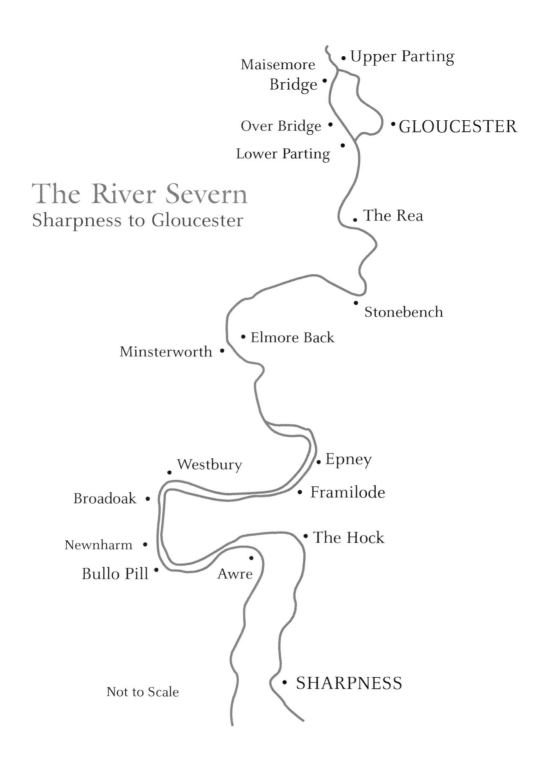

The River Severn
Sharpness to Gloucester

Upper Parting

Maisemore
Bridge

Over Bridge

GLOUCESTER

Lower Parting

The Rea

Stonebench

Elmore Back

Minsterworth

Epney

Westbury

Framilode

Broadoak

The Hock

Newnharm

Bullo Pill

Awre

SHARPNESS

Not to Scale

The River Severn
at Gloucester

To Tewkesbury

Maisernore Weir

Upper Parting

East Channel

Globe Inn

Alney Island

Jolly Waterman

West Channel

Westgate Bridge

Lower parting

Llanthony Weir

GLOUCESTER

A48

A38

Minsterworth

Rea

Weir Green

Stonebench

Canal

Canada. On a very large tide, the difference between low water and high water can reach 15 metres just a few miles above Beachley. From a little above the M48 Severn Bridge down to Avonmouth is the area where the water level drops lower than any other point on the Severn or Bristol Channel on the ebbing tide.

At low water above Sharpness, the river takes on a completely different look to when the tide surges in. It looks inviting to walk out on the sands as they glisten in the sunlight, but be warned: many people have drowned when doing just that. People have been seen walking on the sands unaware that the tide is due. It is too late when they realise the tide is coming in fast and in no time the sandbanks are surrounded by water, cutting them off from the shore. The Severn has the reputation of being the most dangerous and treacherous river in Britain; it has claimed many lives, the majority of which have been lost along this stretch of the Severn from Saniger Sands (off Lydney) to up as far as Newnham on Severn.

Sharpness has been an important port since the opening of the new dock on Wednesday 25 November 1874. On that day, two ships came into the port from the Severn on the 8 a.m. tide on a miserable, wet and stormy day. Today, Sharpness has a thriving trade with ships from mainland Europe bringing in a succession of varied cargoes ranging from fertilisers, coal, grain and cement. It even has an export trade, with many cargoes of scrap steel going to Spain and grain bound for the north of England. Ships bound for the port must have a pilot, who will have joined the ship further down the Bristol Channel off Barry. No pilot is happy when bringing a ship up and into Sharpness until they are tied up in the basin or lock. One of the most difficult manoeuvres is swinging the ship around in the river off Sharpness, especially when there is a large spring tide running. The ship has to be swung 180 degrees around to face downriver while punching (pushing against) the incoming tide. With great skill, the ship has to be brought close to the north pier and slowly come around and in through the narrow entrance into the dock basin. Of course, ships leaving Sharpness have to have a pilot to take them down the estuary and into the Bristol Channel.

Whereas the tide, as it comes in at the seaside, adopts a quite sedate appearance and takes some time to reach high water, in the Severn Estuary it is a different story. Standing at the Old Dock entrance at Sharpness (now the Severn Area Rescue Association (SARA) lifeboat station), watching a large tide racing in at a speed approaching 10 mph is spectacular – no bore wave to be seen, simply the power of the Severn tide. One hour later is high water, and a few minutes afterwards the tide turns and begins to ebb.

Not so many years ago, it was relatively simple to see the bore in more than one location on the same tide. Now, sadly, this is not the case. There are far more vehicles on the highways, all wanting to park somewhere. When parked, there are still the crowds of people who get in the way. Should you be fortunate and

A view of the Severn Estuary from the middle of the river.

A view across the estuary from below Purton to Etloe.

A view of Sharpness docks from the river.

South Pier at Sharpness at low water.

A ship, the *Opus*, bound for Sharpness, swinging off the piers.

The *Opus* gently coming in between the piers at Sharpness.

Low water at the old entrance to Sharpness.

Tide rising at the old entrance.

The Severn Area Rescue Association (SARA) lifeboat station.

SARA lifeboat at speed on the Severn off Sharpness.

surprise yourself by being the only person there to see the bore (it can happen!), enjoy the peace and magic as the bore thunders past.

Obviously, if you are going to view the bore, you can travel along the east bank or the west bank of the Severn.

*   *   *

Let us start our first journey along the Severn shore on its east bank from the marina at Sharpness, where from 1939 until 1967 the Merchant Navy had their training ship TS *Vindicatrix* moored. On the ship and at the nearby camp, young lads fresh from school would spend twelve hard weeks training to be seamen or catering stewards. Discipline was harsh for the 70,000 or so 'Vindi boys' who passed through TS *Vindicatrix* but this was a tough era and men needed to be tough to survive their years at sea.

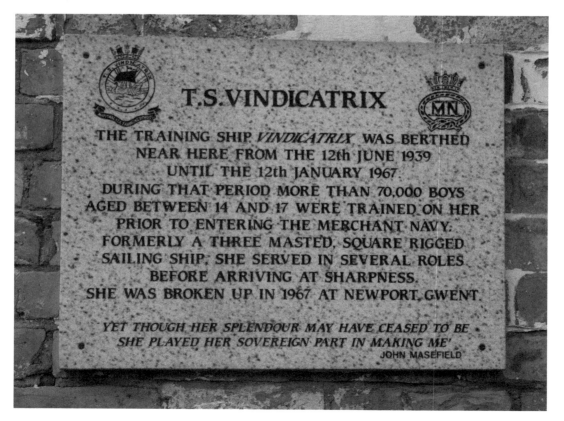

Memorial plaque for TS *Vindicatrix*.

The towpath of the Gloucester and Sharpness canal runs parallel with the Severn from the marina for a couple of miles until reaching the village of Purton. Along the foreshore of the Severn between the towpath and the river lie numerous old barges, which are now known as the 'Purton Hulks'. They were deliberately driven ashore to prevent erosion of the bank and, more importantly, the collapse of the canal bank.

One of the biggest fears for the crew of vessels in the estuary bound for Sharpness is fog, especially when it coincides with a large tide. Out in the river at low water, the wrecks of the two tanker barges *Arkendale H* and *Wastdale H* can be seen, a reminder of the fateful night of 25 October 1960. These two barges with others from Avonmouth and Swansea were laden with oil and making passage up the estuary on the night tide bound for Sharpness. As they approached the port entrance, a thick fog descended over the river and these two barges were swept upriver and collided with the Severn Railway Bridge. Two spans dropped down onto the barges, resulting in an explosion, whereupon the river became ablaze with petrol and heavy oil from the two craft. Each barge carried a crew of four men, and five of these eight were killed.

Beyond Purton is Slimbridge, home of the Wildfowl & Wetlands Trust, which was founded by Sir Peter Scott in 1946. The wetland reserve is there to look after and study ducks, geese and swans from around the world. As it is difficult to access the banks of the Severn here, it would be wise to travel from Purton direct to a location called Splatt Bridge. A landmark to aim for is the Bell Inn at Frampton on Severn. From here, proceed along the road, which passes the largest village green in England, past the parish church of St Mary the Virgin, until you reach Splatt Bridge (one of the sixteen bridges that cross the Gloucester and Sharpness Canal). A small car park is located before the bridge, as you cannot drive over it. From here, stout walking shoes and protective clothing are recommended, as a long walk awaits you.

On the public footpath, head towards the river and aim for Hock Cliff to the right of you. This is the start of the famous horseshoe bend of the Severn and where the tide becomes confused. A word of warning: be careful and don't get too close to the river or the foot of the cliff when the tide is due. As a large spring tide races in, it becomes confused as the waters crash into the rocks at the foot of Hock Cliff. In a state of turmoil, the tide sorts itself out and turns sharp left to head across the river towards Bullo Pill.

It is advisable to return to your car and drive to Arlingham Passage at the very apex of the horseshoe bend. On the opposite bank is the village of Newnham on Severn from where in 1171 King Henry II sailed to invade Ireland. At Arlingham Passage, although desolate, there is a seat conveniently placed to sit and enjoy the view across the quarter-mile width of the river. There is not only an attractive view, but also the pleasant sound of the bells from the parish church sited on the hill above Newnham on Severn.

One of the many 'Purton Hulks'.

Another hulk at Purton.

A weather-beaten hulk.

The wrecks of *Arkendale H* and *Wastdale H.*

Across the Severn from Arlingham to Newnham on Severn.

Arlingham is where a neap tide finishes, but not the spring tide. That arrives with great power, forcing a reversal of the natural current of the river, which will be seen 25 miles away at Tewkesbury. Now we have a bore, albeit small, yet the excitement builds as the wave forms and rapidly covers the sands.

Through narrow country lanes we now make our way to Framilode, from where there is a wonderful view of the Forest of Dean far in the distance. On the opposite bank is Rodley, with the sands (named after the village) glistening in the sunshine. Numerous gulls remain on these sands right up until the tide appears, no doubt annoyed to have their lifestyle interrupted each day. It's still not the best wave to be seen, but from here the banks of the river begin to narrow, which is the secret of that large bore wave.

Framilode has a certain magic – it is somewhere that time has forgotten. The church of St Peter, built in 1854, stands close to the Severn and, until a high concrete floodbank was built in 1961, it used to suffer from the effects of a high tide. Close by there used to be the entrance to the Stroudwater Canal as it joined the Severn via the now disused lock. A little up the lane are a row of delightful cottages that line the disused canal.

A little way up the river is the Anchor Inn at Epney, a favourite watering hole today as it was many years ago. You will not have failed to notice the Gloucester and Sharpness Canal as you travelled to the banks of the Severn

An inflatable near the bank races ahead of the bore as it approaches Framilode.

The bore approaching Framilode

Gulls flee from the bore near Framilode.

Turbulent water at Framilode following the bore.

30

Sunset over the Severn at Framilode.

The church of St Peter at Framilode.

at Epney. Opened in 1827, it was built to avoid the treacherous and difficult-to-navigate part of the Severn between Gloucester and Sharpness. Prior to the canal opening, it was possible for sailing craft to travel to Gloucester from the Bristol Channel on one tide. However, this was not the case when returning with the ebb tide. It could take two or even three tides to make the passage back to the sea; thus, where the small ships would lie between tides is where inns were built. One favourite place that vessels would lie was at Stonebench, where the long derelict inn was recently demolished.

Between Epney and Stonebench is a public footpath that is well worth a walk, even without the bore. It passes Longney with the wide expanse of the sands over on the west side of the river with the deep-water channel hugging the outside of the bend.

Continue walking along the footpath, for ahead of you is Elmore Back and at last the bore becomes something worth seeing. This area along the Severn can be a nightmare to find. Roads seem to cross each other and although well signposted this area is still a maze. It may be wise to mention that it can prove difficult to travel around here in a car whenever a good bore is forecast in the local press.

As you drive along the back lanes from Longney towards Elmore, you will come across the turning to Elmore Back. Treat this road with care, as it is a mile in distance to the hamlet and alongside the Severn with very few passing places. This used to be an excellent place to see the bore, but sadly this is no longer the case. The first problem is finding a place to park the car, then finding an easy way up onto the flood defence bank that runs alongside the river.

A far better option is to drive further along the lane to Elmore and take the single-track lane down to Weir Green. It is only a short lane and if you are feeling energetic you can leave the car at the junction and walk down to the river. There are only a few spaces to park by the river, so you can imagine the chaos there when a good bore is predicted. Having reached the river, turn left and walk a little way downstream when suddenly you will be faced with an open expanse of river with views for some considerable way in the distance. With the sun rising from the east, this is a good place from which to take photographs, as the sun will be behind you.

From Elmore there is a decent road to Stonebench. However, be warned: in times of media attention about the bore, this area is a nightmare. Parking is limited and motorists will insist on leaving their cars at the side of the road, making it difficult for others to pass. Even more worrying is that the road is lower than the top of the bank of the Severn, which runs parallel to the road with only a narrow strip of land separating the two. It has been known for water from the Severn to find its way through and over the bank and to flood the lane. Within the last couple of years, the fire service were called to rescue a lady from her car when water from the river began to enter the car and trapped her inside.

Cattle on the banks of the Severn at Elmore Back.

Cattle on the banks of the Severn between Weir Green and Elmore Back

The River Severn at Weir Green.

An orchard on the banks of the Severn at Weir Green.

Stonebench was always rated as the best place to see the bore and still is, if only you could see it. Now the numerous trees that line the bank prevent a good view of the river and with the added problem of parking and access perhaps this is an area best avoided. Many place names along the course of the Severn have the same endings, which are associated with the river. Names ending in 'lode' mean a shallow place in which to cross, as in Wainlode, Hampton Loade, Upper and Lower Lode. Likewise, names ending in 'bench' refer to a shelf of rock in the bed of the river. Here at Stonebench a line of rock used to extend three quarters of the way across the river until it was removed some years ago. This was one of the reasons craft could not navigate all the way down the Severn from Gloucester on the ebbing tide. As the level of the river dropped, craft were unable to pass over the rock at low water. It wasn't unusual for them to be stranded here for anything between a few hours and a week or more. As the men became stranded on their craft, they needed somewhere to socialise, so inns were built at these shallow locations. The line of stone in the river did help make the bore at Stonebench more spectacular than elsewhere on the Severn. Since the Severn Catchment Board removed all stone from the bed of the river in the 1930s, the river between Minsterworth and Gloucester has an even flow, making the Severn bore more predictable in how it will perform.

In the village of Hempsted at Gloucester is a small lane leading to Upper Rea. A public footpath will take you back in the direction of Stonebench and again this was once a good spot from which to see the bore. Not many people were aware of this footpath and bore watchers were usually there by themselves, away from all the crowds, in peace and isolation. That is still the case, but visibility can be restricted because of the great amount of foliage along this bank.

A footpath runs parallel to the Severn from Upper Rea towards the Lower Parting at Gloucester. This is not the most pleasant of walks, as it follows the riverside perimeter fence of a large landfill site. Refuse lorries are continually dumping rubbish from our cities while thousands of gulls endeavour to feed on the food waste. Such an unattractive location means that very few people come here to see the bore. Towards the corner of the Lower Parting used to be an ideal spot to view the bore and see how it divides as water enters the east channel. Now that the bank is so overgrown, it is impossible to get a good view.

Although it means another long walk, it is possible to reach the opposite side of the east channel and there are clear views looking down the river to see the bore approaching. Access to this location is either from Telford's historic Over Bridge or via the official tourist trail from Castlemeads car park.

A little way below Gloucester, the river divides into two and this location is known as the Lower Parting. As the bore passes the split at Lower Parting, it loses some of its force as part of the wave travels up the east channel to Llanthony Weir. Three miles further on upstream, the river becomes one again at – would

The Severn Bore arrives at Weir Green.

Spectators at Weir Green, disappointed by an underwhelming bore.

Spectators watch the river level rise after the bore has passed at Weir Green.

A couple photograph the bore near Stonebench as it is ridden by a kayaker and jet skier.

A grandfather and his grandchildren watch the bore as it passes Stonebench.

An old retired barge moored at Stonebench at the peak of the tide.

The stone that was removed from the bed of the river and gave the location the name of Stonebench. This stone lies alongside the public footpath.

A postcard of the Severn Bore at Lower Parting on 4 September 1921.

The bore racing into the east channel of the Lower Parting.

Surfers being filmed from an inflatable as they ride the bore at the Lower Parting.

A lone canoeist dodges debris left by the bore at the Lower Parting.

you believe – the Upper Parting. Both channels are known as the Parting. The east channel runs over Llanthony Weir, past Gloucester Lock and along The Quay towards Sandhurst and is the navigational channel. The other leg of the river is the west channel and runs under the three bridges: Over Railway Bridge, Telford's old Over Bridge (now redundant and purely a historic monument) and the relatively new A40 Over Road Bridge. Many hundreds of people crowd onto Telford's Over Bridge to see the bore, yet this has to be the worst place to see it. The view of the approaching bore is spoilt by Over Railway Bridge and from this height the bore looks insignificant.

The river continues its course to Maisemore Bridge, where finally the bore ceases as it hits Maisemore Weir.

Telford's historic Over Bridge.

Over Railway Bridge.

We start our second journey along the Severn shore on its west bank, 14 miles from Gloucester on the A48 at Bullo, and follow the public footpath to below Bullo Dock. Here you can patiently wait to see the tide coming across the river from the direction of Hock Cliff (Frampton on Severn) after it has sorted itself out. There's not much of a wave as it passes, and in some sense it will seem a waste of time after it has gone by. The interesting part is seeing the turmoil across at Hock Cliff as the incoming tide hits the rocks and returns on itself before realising it should have turned left and continued across the river towards Bullo.

Driving along the A48, the village of Newnham on Severn is soon reached. A visit to the parish church of St Peter's at the top of the hill is worthwhile. The best view of the bore is rarely from above, but here there is a different perspective as the Severn opens out below you. Before the tide is due, a large expanse of sand can be seen. Watch as the leading wave sweeps up the river and begins to fill and cover the sand.

The car park on the Gloucester side of the village, alongside the bank of the Severn, is well provided (but you need to be there early). It also provides a good viewing point for the bore, but not if you wish to see a wave of some size. For that, remain patient until you see the bore further upriver, because it is possible to see the bore at Newnham then motor further up the river and see it again. Indeed, you could see it several times at different locations, but one thing may prevent this: the mass of people who come to see the bore. Parking is the problem; thus many people arrive a while before the bore is due simply to secure somewhere to park.

Remaining on the west side of the Severn, we journey on to Garden Cliff at Westbury on Severn. Again there is a choice: either climb up to the top of the cliff, where the bore can be seen approaching from some way downriver, or remain at river level at the Strand. On no account be tempted to walk along the foot of the cliff when the tide is due as you will be in danger of being cut off. The river is wide here, with a vast expanse of sand over towards the east bank, known as Pimlico Sands. The tide sweeps around the west shore from Broadoak, seeking the deep water alongside the sands. At times with a large spring tide, the leading wave will break over the stonework placed along the riverbank, adding to the excitement.

Sadly, there are not too many suitable places to see the bore between Westbury on Severn and Minsterworth. Further along the A48 towards Gloucester is the Severn Bore Inn at Minsterworth, which is well worth a stop, especially if they are having one of their features of breakfast and bore incorporated. Although this is not the best place to view the bore, it is very civilised with refreshments on hand.

Still in Minsterworth, a wise first-time bore watcher will travel down Church Lane and park at the church car park. Then it is only a short walk to the riverbank

The view across the river of the spire of the church at Westbury on Severn.

Garden Cliff at Westbury on Severn.

The view from the top of Garden Cliff, looking downstream.

Looking upriver from Garden Cliff.

to wait for the bore, which can be seen approaching in the distance. While you wait patiently, ponder on this thought. Here at Minsterworth the bore arrives at the same time as it is high water at Sharpness. The tide has come in, and as it travels up the Severn it is pushing back the current that sends water down from the source of the river. In simple terms, it is reversing the flow of the river. One of the strange phenomena of the bore is that as it passes Minsterworth the tide is beginning to ebb at Sharpness and continues to do so as the bore races towards Gloucester.

Minsterworth is a very popular area for watching the bore, so there is a need to arrive early – very early, in fact, if you want to get parked and avoid the chaos on the local roads. It is possible to get away from the crowds by continuing to walk upstream from the church for as far as you want.

A few miles above Minsterworth, Rob Keene of Over Farm Market offers exciting rides chasing the bore in a trailer towed by a tractor in fields along the banks of the Severn.

Nigh on two hundred years ago, a series of weirs were built across the Severn between Gloucester and Stourport. These have the effect of raising the level of the river to enable large barges to navigate as far inland as Stourport. Two of these weirs are at Gloucester: Llanthony Weir in the east channel and Maisemore Weir in the west channel. The River Severn at Gloucester is quite complicated, as mentioned earlier with regard to The Parting. An added complication is the fortnightly spring tides as the river splits at the Lower Parting.

A third of the tide goes into the east channel while the remainder rushes on towards Maisemore. The bore is stopped as it hits both weirs, the wave rolling back for a short distance before hitting the weirs again. The continuing force of the water rises quickly until it tops the weirs and rushes on up both channels without a bore wave. In effect, both Llanthony and Maisemore weirs are the end of the bore but not the end of the excitement.

The force of the tide continues to reverse the flow of current along the east channel as the tide in the west channel passing Over acts in a similar way. As the water in the west channel reaches the Upper Parting, a large whirlpool is created outside in the wide channel of the Severn. Part of the reason for this is that a third of the water coming out of the west channel swings and travels down the east channel with the current towards the White Horse Chinese restaurant (previously the Globe Inn) at Sandhurst.

Here the current meets the surge of water of the tide, which is travelling at speed from Gloucester. The river is in turmoil for a short while as the two converging waters sort themselves out and of course the tide wins. Soon all the water is rushing back up to the entrance to the Upper Parting and out into the main body of the River Severn. All the while, it is the reversal of the current

Crowds watching the bore approaching at Minsterworth.

The bore along the bank at Minsterworth church.

Riding the bore at Minsterworth.

The bore cuts through the early morning mist at Minsterworth.

Trees along the bank at Minsterworth. When a good bore is predicted, arrive early to secure the best viewpoint.

Spectators of all ages crowd the bank at Minsterworth in eager expectation of a good bore.

A Sky News presenter reporting on the bank of the Severn at Minsterworth.

BBC Radio Gloucestershire interviews bore watchers after the bore has passed Minsterworth.

which is unique, with the level of the river rising fast and flowing at speed back to Tewkesbury.

At Tewkesbury there is another weir at Upper Lode with a large lock for navigation built alongside. The weir stops most of the flow of water but a large spring tide will rise above the weir to further effect the flow above.

Before the weirs were built, the effect of the bore was felt as far inland as Worcester. It is not unknown for both top and bottom gates at Upper Lode Lock to be open together as the water level becomes equal on both sides of the weir.

Quite a considerable amount of debris appears in the Severn during the course of the spring tides above and below Gloucester. As each tide arrives, it moves items like oil drums, gas canisters, parts of trees (even complete trees), dead farm animals, indeed anything that floats. Debris travels upriver with the tide and travels back down on the ebb, becoming stuck on the bank or sandbank as the water runs out. This is repeated twelve hours later and each day until the cycle of spring tides has ended. The debris then remains on a sandbank until, two weeks later, it all moves again. Very rarely is it taken down into the estuary and out to sea. Please don't get the idea that the Severn is a dirty river, far from it: the river is one of the cleanest in Britain. It has to be, for over 6 million people use water from the Severn each day as it is pumped out, cleaned, used and pumped back in again. What makes the Severn look dirty at Gloucester with its brown, muddy appearance is the silt it has carried on the tide from the estuary. Take a jar, fill it with this brown water and wait a while. Soon the sand will settle on the bottom of the jar, leaving crystal clear water above.

The most stunning view of the River Severn is from the most peaceful area of Gloucestershire with the romantic name of Pleasant Stile. Part of the enjoyment of being there for me is how I stumbled across the view by accident. So look on a map and head for Littledean. From Pleasant Stile on a clear day, Gloucester can be seen clearly in the distance as well as a view of the Severn from Epney down around the great horseshoe bend past Westbury on Severn, Broadoak, Newnham on Severn, across to Hock Cliff at Frampton on Severn and down to a little above Sharpness.

To view the bore, a pair of sunglasses would be advisable – and hope for clear visibility. The experience of seeing the bore from this great height is remarkable and something very few people would know about.

So is that the bore in a nutshell? Not quite, for there is so much else associated with spring tides throughout the year: wildlife, people's enjoyment and transport all have a part to play.

The power of the tide, forcing the current back at Upper Rea, with much debris in the water.

The power of the tide against a tree stump in the Severn.

An inflatable boat rides the bore at Weir Green, while much debris has accumulated on the right bank.

More debris from a previous bore at Weir Green.

The horseshoe bend in the River Severn, as seen from Pleasant Stile near Littledean.

The Severn's horseshoe bend is the largest, most inwardly set, geographical feature of its type in lowland Europe.

Another part of the horseshoe bend in the River Severn.

A close-up of part of the Severn's horseshoe bend.

Another view of the Severn from Pleasant Stile.

Pleasant Stile offers some of the finest views of the Severn Vale.

When to see the Bore

It is your decision to decide where and at what time to see the bore – whether to see a daytime or night bore. To assist you in making up your mind, there are several factors to be taken into consideration. The following are the golden rules to obey to make your trip to see and enjoy the bore successful.

Location
- Choose your location carefully.
- Is there parking available?
- Is there a spot from which to clearly see the bore?
- If necessary, check the area out a few days before.

Time
- Make sure you have worked out the bore's time of arrival accurately.
- Have you added the hour when converting from GMT to BST?
- Arrive early – an hour early if necessary.
- Usually the bore is late, quite often twenty minutes late. Sometimes it is on time, but rarely early. Unless you allow yourself plenty of time to get parked and walk to a suitable location, you could miss it.
- Always remember one of life's magic quotes: 'Time and tide wait for no man.'

Safety
- Have you considered your safety? A large tide can soon top the bank and knock you off your feet if you happen to be stood too far down a bank near the water's edge. Some people are stupid enough to do this.

Don't be disappointed if the bore is not as good as you thought it would be. There are very few bores that live up to people's expectations. Stay awhile to see the river rising as the great mass of water that follows the bore speeds by.

You may be tempted to walk away from the riverbank saying, 'What was all that about?' If this should be the case, stop and remember that what you have just seen is one of nature's great phenomena. Nothing man has done has influenced the bore: it is something that has been happening ever since the River Severn ran down to the Bristol Channel and out to the Atlantic. The Romans, when they first came to this area, treated it with caution; indeed, some say they were frightened of it. One thing man can do is destroy the bore, as would be the case if ever a decision is made to harness the power of the estuary by building a massive concrete dam across the Bristol Channel from Cardiff to Brean. This idea has been talked about ever since electricity was discovered. Now that fossil fuels are running out, alternative ways of generating power have to be found.

It is not easy to explain how the tidal cycles of the Severn create a bore. There are many misleading stories about how many bores there will be and when the bore will occur in the course of a year. Hopefully, the following explanation will settle some of those arguments and you will visit the banks of the Severn with enough knowledge to appreciate what is happening before you.

Each year there are between 250 and 260 bores, including large and medium ones. As there are two tides each day, this reduces the figure to a bore appearing on 130 days of the year. To confuse matters a little, there are another 150 tides each year that have the surge effect of a bore but really comprise a long, smooth wave that does not immediately reverse the flow of the current.

As mentioned previously, bores arrive on a fortnightly cycle and within the fortnight the bore can be seen for three, four or even seven days. The fortnightly span is taken from the middle day of one bore cycle up to the middle day of the next cycle, thus half a lunar month. As can be seen from the monthly graph, the bore grows in size from the first predicted day to a maximum on the middle day and then declines. In each lunar month, there is the new moon and the full moon with each producing a spring tide. For seven consecutive months the new moon produces large spring tides and for the next seven months the full moon produces the same. This cycle will last for nine years before starting all over again, thus revealing why tide times and height can be predicted so far in advance. This is explained when consulting the monthly graph of tides for Sharpness: during one fortnightly cycle the tides are smaller in height than the other fortnightly cycle.

The number of bores can be broken down even further. There are approximately fifty large bores in one year, and these are the ones that attract the most publicity. To break it down even further, half of these occur at night with the other half obviously occurring during the morning. What's more, these large bores occur during two quarters of the year: February to April and August to October, the Vernal and Autumnal Equinoxes respectively.

It is all a prediction, and in so many cases the bore does not live up to expectation. You may arrive at your preferred location on the day of the highest tide of the fortnightly cycle to be told that yesterday there was a very good bore

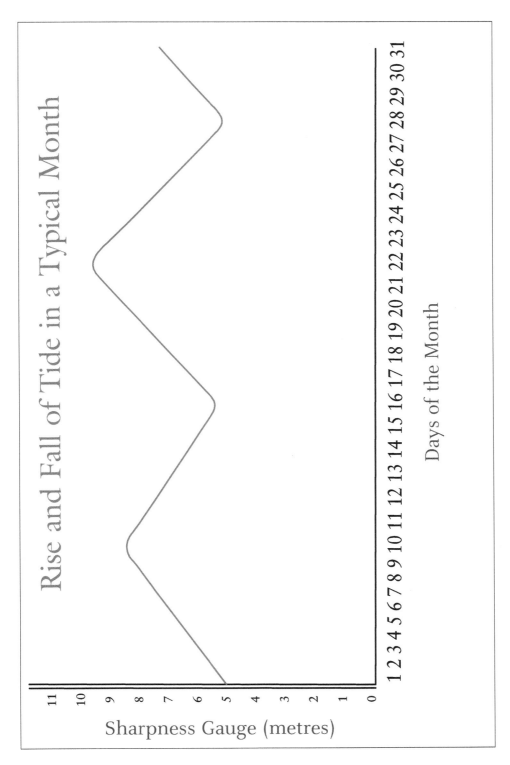

A graph of the fortnightly tides cycle of the Severn Bore at Sharpness.

A block of concrete that was lifted out of the river by the force of the bore and deposited on the bank.

– so good that it lifted a large concrete block out of the riverbed and onto the top of the bank. Intrigued, you go to inspect this cube of concrete and try in vain to move it. You become excited as you are told that conditions are right for an even better bore today.

How wrong could that prediction be? For when it arrived, it was abysmal, so poor that even the surfers could not ride it. This brings home to you that it is only a prediction and yes, the tide had arrived (which hopefully it does, for if it doesn't we are in trouble) but, for whatever reason, conditions weren't right that day for a good bore. Early writers about the bore all state how temperamental it is.

Today, in this modern world of computers and the Internet, it is relatively easy to find out when bores are due. Some sites are better than others but www.severntales.co.uk gives all the dates and times a year in advance, as well as predictions, facts and photographs. If no computer is available, a valuable asset is *Arrowsmith's Bristol Channel Tide Table*, where full details of the bore are given. Another valuable publication is the free leaflet, *The Severn Bore*

Spectators waiting for a night bore at Minsterworth.

Spectators at Minsterworth eagerly await a bore on a bright and beautiful day.

and Trent Aegir, produced annually by the Environment Agency, which may be obtained from British Waterways and other tourist centres.

All the large bores occur between 7 a.m. and noon, and 7 p.m. and midnight; thus, you will have to consult the official tide times and predictions from the sources previously mentioned to plan your visit accordingly. If you choose to view the bore in the morning and then decide to view it again, but this time at night, you will be in for a treat.

Imagine a cold yet clear autumn night with a bright new moon shining overhead and the moonlight sparkling on the waters of the Severn as the current slowly travels down from Gloucester. The night is still and extremely quiet – not a sound from any creature, not even any wind to rustle the few remaining leaves on the trees. When suddenly, all this peace is shattered as a noise like an express train approaches. The branches of the trees lining the banks of the river snap in a frenzied fashion and then all hell breaks loose.

Although dark, the moonlight illuminates a giant wave as it quickly surges by. A serene setting now becomes a frightening yet exciting moment: water rising at your feet, with the noise of breaking branches replaced by the flow of water being reversed as the tide pushes against the downward current of the river.

This is the Severn bore at night – quite different to a daytime bore, but becoming increasingly popular with people wanting to see this amazing phenomenon in a different light.

Interesting Facts about the River Severn and the Bore

The Severn Bore occurs throughout the year, with the largest bore appearing during an equinox. These happen twice each year: February to April, and August to October.

The time of a bore will be between 7 a.m. and noon and again between 7 p.m. and midnight. A precise time will depend on the location from where you choose to view the bore. The largest tides occur at Sharpness at about 9 a.m. (GMT).

To further confuse matters, larger bores occur one to three days after a new moon and a full moon, while smaller bores happen on days preceding and following the maximum height of the tide.

A tide of 8 metres and above, as showing on the gauge at Sharpness Docks, will produce a Severn Bore. A tide of over 9.5 metres will give a large bore.

Lower down the Severn Estuary the speed of the bore will be between 8 and 13 kilometres per hour (5–8 mph) and above Minsterworth between 16 and 21 kilometres per hour (10–13 mph).

A large spring tide, as it passes Sharpness, is flowing at a speed of approximately 9.5 kilometres per hour (6 mph) and the same as it ebbs back out. From Minsterworth to Gloucester this same tide will be travelling at about 19 kilometres per hour (12 mph).

The height of a very good bore can be about 3 metres (10 feet) in midstream, although a greater height can be seen as it runs along the bank of the river. However, most bores are classed as good if they pass with a height of 1 metre (3 feet).

The width of the bore really depends on where it is in the river. Along the estuary, it can be 250 metres (820 feet) wide, whereas between Minsterworth and Gloucester it narrows to between 80 and 95 metres (262 and 312 feet).

There are about 260 bores per year, which with two per day means that for 130 days of the year there will be a bore. For 25 days there should be large bores both during the morning and evening but this will very much depend on conditions in the river.

A good bore wave will run for between 2 hours and 10 minutes up to 2 hours and 35 minutes. Again, this depends on river conditions.

According to modern records, the largest bore was on 15 October 1966 when a height of 2.8 metres (9.2 feet) was reached as it passed between Weir Green and Stonebench.

The chart below indicates the time of the bore at popular viewing places, based on High Water at Sharpness:

Awre	subtract 1 hour 30 minutes
Hock Cliff	subtract 1 hour 25 minutes
Bullo Pill	subtract 1 hour 7 minutes
Newnham on Severn	subtract 1 hour
Broadoak	subtract 50 minutes
Westbury on Severn	subtract 45 minutes
Framilode	subtract 25 minutes
Epney	subtract 20 minutes
Severn Bore Inn	subtract 7 minutes
Minsterworth	high water at sharpness: 0 minutes
Elmore Back	high water at sharpness: 0 minutes
Stonebench	add 15 minutes
Upper Rea at Hempsted	add 20 minutes
Lower Parting	add 33 minutes
Over Bridge	add 35 minutes
Maisemore Bridge	add 40 minutes

Bore	Barometer Reading	Winds	Freshwater in River Severn	Channel through Estuary Sands
Will arrive later if:	High	Strong N to E	No freshwater	Bore cutting a longer channel through estuary sands
Height smaller if:	High	Strong N to E	No freshwater or too much freshwater	
Will arrive earlier if:	Low	Strong S to SW	Up to 5 metres (16 feet) of freshwater	Bore cutting a shorter channel through estuary sands
Height increased if:	Low	Strong W to SW	0.75 metres (2.5 feet) of freshwater in river below Gloucester	Channels through estuary sand are well scoured

In addition to the factors above, there obviously has to be a spring tide. These are listed in tide tables, on the Internet, and sometimes in the local press.

Please note that the term 'freshwater' refers to the amount of water in the Severn above the normal seasonal level. Rainfall farther up the river will cause the level of the Severn to rise.

The bore through the trees at the Lower Parting.

A clearer view and calmer waters at the Lower Parting, after the bore has passed.

Surfing the Severn

Surfers – you either love them or loathe them! You will always see at least one surfer riding the bore wave, but usually there are a bunch of them, all trying to stay with the wave as far as they can. Then there are the large inflatable boats with noisy engines, some in front, some behind the wave, ready to retrieve their mates from the water when they lose the bore.

It is generally accepted that the first man to surf the Severn was Colonel Jack Churchill, who on 21 July 1955 at 10.30 a.m. dropped himself into the river a little below Stonebench, ready to surf the bore. Tidal predictions were good and he was excited when he saw a good bore approaching him from downstream. As the leading slope of the bore was beneath him, he deftly balanced himself on his board and began to plane forward with the wave.

He had come to the Severn with plans of a long ride on his board. However, his hopes were soon dashed as the wave ran into shallow water at Stonebench and he was thrown off balance and into the water. His attempt inspired others to try to tame the wave and enjoy the difficulties of surfing the Severn.

It was not until the 1960s that surfing the Severn Bore became popular, and it has increased in popularity ever since. With ease of travel around the globe, surfers come from Australia and America to enjoy the experience. Surfing a wave in a river is very different to surfing a wave at sea. At sea, the experience will only last a very short time, while on the Severn a surfer could surf for up to seven miles!

There are dangers from surfing both in the sea and on the Severn. At sea, surfers face one danger, a rip current (usually referred to as a 'rip tide'), where swimmers are dragged away from the beach, become exhausted fighting it and drown. On the Severn, surfers face a number of dangers. Their fear is being thrown by the wave into the bank and being rolled over at least a dozen times before being released. Battered and bruised, they have the added fear of being struck by large overhanging branches or smashed into the sizeable boulders that line the bank. As already mentioned, there is a significant amount of debris in the river during the course of the spring tides. This poses a problem to surfers. Stories tell of one surfer who ended up with a refrigerator on his board. Besides

A surfer riding the bore as Stonebench

Aboard a boat of surfers, on their way to ride the bore.

An inflatable races to catch the bore wave.

The wake left by a bore-seeking boat on the Severn..

Dave Lawson (left) and a Frenchman surf the bore together.

A host of surfers, including Dave Lawson, ride the bore past Minsterworth.

Dave Lawson and a group of Frenchmen ride the bore on their surfboards.

Dave Lawson, in his signature surfing pose with his hands behind his back, rides the bore at Stonebench.

Steve King (left) on his board.

Russell Winter (left) performing with Steve King looking on.

Surfing the bore near the bank can be dangerous.

Surfers try to keep their balance as the bore propels them up the Severn.

An aerial view of surfers riding the bore.

Crowds come to see the bore, and to cheer on the surfers who brave it.

surfing out on the bore, kayaks and even jet skis are popular, all getting in each other's way at times.

The most experienced surfers treat the Severn with great respect and know how to get themselves out of danger. Their advice is that once you are off your board and at the mercy of the great wave, to swim out to the middle of the river until things calm down. That is another reason for the inflatable boats accompanying the surfers: to help when they get into danger.

There is great rivalry between the numerous surfers out on the Severn; yet when things go wrong, they are there to help each other. Dave Lawson, an experienced surfer, relates how, during his attempt at a world record, he abandoned it to assist with helping a man in a kayak who was unable to right himself and was trapped under the water. Before surfing became popular on the Severn, Dave Lawson and Steve King would have the river to themselves, and although a healthy rivalry built up between them as each tried for the world record, there has remained mutual respect for each other.

Some find it hard to believe, but on different occasions the world record for surfing a river bore has been achieved by surfing the Severn Bore. Dave Lawson first held it in 1988, when he surfed a distance of 4.3 kilometres (2.7 miles). Then in 1996 his great rival, Steve King, doubled the distance and claimed the record. He wasn't to hold it for long, as a few months later Dave Lawson managed to surf the bore from Weir Green for a distance of 9.17 kilometres (5.7 miles).

On 30 March 2006, Steve King entered with his board the waters of the Severn at Collow Pill – a small hamlet a little way below Newnham on Severn – ready to catch the bore and surf as far as he could. Although the largest tide was not forecast that day, Steve knew that river conditions were right to produce a suitable bore to surf for a long distance. The river was crowded with fellow surfers, estimated at between fifty and sixty people, all with their boards, waiting in the Severn for the tide.

Despite the crowd around him, Steve managed to surf a total of 14.9 kilometres (9.5 miles) along the Severn to Bollow near Minsterworth. For 12.2 kilometres (7.6 miles) he was stood on his board and for 2.7 kilometres (1.65 miles) he was lying down. Steve had to wait almost five years until Christmas 2010 to receive official notification that he holds the world record for the longest surfing ride on a river bore and that detail of his achievement will be entered in the Guinness Book of Records.

There is a lot of planning to be done before someone can drop themselves into the Severn and try for the world record. The British Surfing Association has to send an official adjudicator to watch and confirm that the record has been broken and to have it entered in the record books. Whenever Dave Lawson attempted to break the record, he insisted on having a rescue boat follow him. Dave and Steve know the behaviour of the bore more than most and agree that from Weir Green to Maisemore Weir is the longest distance the wave travels without breaking up.

Dave has a particular stance while surfing the Severn on his board. He stands in an unusual way with his hands behind his back. Whenever asked why he stands like this, his answer is, 'It's somewhere to put my hands!' With many surfers out on the Severn at any one time, some can get in the way of others. If either Steve King or Dave Lawson is going for a new record, they only have to ask others to keep out of their way and they do so.

With all the debris carried by the river on the spring tides, surfers tell tales of what items have ended up on their boards – anything from tyres to canisters, and as said before, even a refrigerator once ended up on Dave Lawson's board.

With eighty rivers around the world producing good bores, it is no wonder that surfing has become very popular on them. Most surfers would love to surf every one, but there are obstacles in the way on some. None more than where the greatest bore occurs: the Qiantang River in China, known locally as the Silver Dragon. Here a bore wave of 9 metres high (30 feet) travels at 25 miles per hour along the river, which is most menacing to shipping. Following the death of twelve spectators a few years ago while watching a night bore at Hangzhou, it is now forbidden to surf the bore. This is not surprising when considering that during the past twenty years over a hundred people have lost their lives on this great river. (On the Internet, YouTube has footage of the Qiantang bore with people being washed into the river.) The law does get broken, mostly by foreigners who can't resist the temptation of surfing the largest bore in the world.

The Qiantang River is very wide, but also quite shallow, with the bore breaking as a continuous wall of white water. In the deeper parts of the river, the bore can be ridden at the edges, which makes for safer surfing. Unusually, this river always has a bore wave somewhere along its estuary on each tide, unlike any other river in the world.

No matter if the surfers are out on the Severn to compete for another world record or simply to enjoy themselves, it is without doubt one of the most difficult challenges in the world. It is no place for surfers who have never attempted it before without seeking help and advice from the more experienced, such as Dave Lawson, Steve King and others. Staying on the board must be the easy bit – avoiding the debris, rocks and overhanging trees, the most dangerous.

Whether they stay on their board for only a few minutes or manage to surf with the bore for a few miles, surfers all say how wonderful the experience is and how the rush of adrenalin spurs them on to keep surfing the Severn and other rivers around the world. When asked how it compares with surfing the sea, most answer that they have never tried it! So it would appear that there are two distinct types of surfers around the world.

Fishing in the Severn

Elvers

Turn the clock back a few years and the people of Tredworth, a small suburb of Gloucester, could buy elvers by the pint. Indeed, the price of a pint of elvers only cost pennies, whereas today a single elver can cost a fortune. The elderly people of Tredworth still talk about the delicacy of eating a plate of elvers; some swear that they are an aphrodisiac. Others turn their noses up at the thought of eating these strange creatures, for it takes a strong stomach to want to eat them.

Not many miles from Gloucester is the delightful village of Frampton on Severn, where each year the annual fair was held. During this weekend, an elver-eating contest would be held where long tables were laid out on the village green and contestants would sit ready to gorge themselves. With plates in front of them, forks at the ready, the elvers, all freshly cooked, would be equalled out and on the given signal mouths would be filled. The crowds watching this unpleasant spectacle would pull faces in horror as each contestant shovelled the elvers into their mouths, determined to be the champion.

Sadly elvers can no longer be bought on the streets of Gloucester, and the elver-eating contest at Frampton had to stop due to the lack of elvers, which was made worse by the fact that the cost of elvers suddenly soared when entrepreneurs realised the potential of making lots of money by exporting elvers to other European countries and to Japan.

In turn, the elver fishermen could make a lot more money by selling their catch to an elver station rather than going around the streets of Gloucester with a bucket of elvers to sell to customers in the pubs. These were the innocent days of seeing local men, with strange nets on the roof of their cars, heading for the Severn to stand on the bank with Tilley lamp lit, waiting for the tide to turn before getting to work with the net.

What is this creature we call an elver? For centuries, knowledge of the elver and eel remained a mystery until, early in the twentieth century, expeditions were made to the Mediterranean and North Atlantic to conduct a study about them. It was discovered that larvae of the European eel travelled with the Gulf Stream from the Sargasso Sea, south of Bermuda, across two thousand miles of ocean, and after three years they reached the coast of Western Europe. As they approach our shores, the elver – or to give its common name, the glass eel – has grown to a size of 75 to 90 mm (2.9 to 3.5 inches) and looks very much like a transparent worm.

Their aim in the UK is to travel far up the River Severn into fresh water, but they have many obstacles to overcome before reaching their destination. They swim in great shoals as they arrive in the Severn Estuary from about early February through to April. Usually the 'elver season' begins during the large spring tides of February – a time for the fishermen to eagerly await the first catch of the season.

The clever elver fisherman will study the tide, listen to stories of where the elvers were seen on previous days and choose his spot on the Severn accordingly. He would have earmarked his 'tump' (the spot on the riverbank where he will stand to fish) and will patiently wait for the tide. The big catches are caught at night. Thus, a few years ago it was common to see the banks of the Severn lit by paraffin-fuelled Tilley lamps. Today, most men wear the ever-popular headlamp strapped to the head.

As the tide approaches, a shout will be heard saying 'Tide-oh', then more patience is required while waiting for the tide to turn. As the tide in the river ebbs, the elver fisherman will dip his strange-looking net into the water hoping to secure a big catch. Should he be lucky enough to catch some elvers, he has to treat them with care as he tips the contents of the net into a bucket he has positioned close by.

His net is usually handmade with a frame of willow or aluminium tubing to a size of 1 metre (3.3 feet) long, 0.6 metres (2 feet) wide, and 0.5 metres (1.6 feet) deep. The net is tightly stretched over the frame and is a form of cheesecloth. With a 1.5-metre (4.9-feet) pole attached to one end of the frame, he is ready to dip his net into the Severn.

The nets are dipped into the river, lifted out and the catch tipped into the bucket. This action is repeated over and over again for at least two hours. Each lift of the net is termed a *shutt* and the amount of the catch is measured in pints (1 pint equals 0.56 litres). With these strange, translucent creatures squirming to escape up the side of the bucket, a few inches of foam is formed, giving the appearance of a freshly drawn pint of beer. The elvers must not remain in the bucket for too long for fear of damaging them, so regularly and carefully they are placed on the hessian base of the numerous trays that accompany each elverman.

Elvers swim with the tide to travel further inland. Then, when the tide turns to ebb, the elvers have to swim against the flow to continue with their efforts to swim upstream. These wriggling elvers come close to the bank, which location affords the least resistance when swimming against the ebb. This is where they are at their most vulnerable as they fall prey to men with their large nets. Prior to the 1980s, elver fishermen could command a figure of up to £700 per kilogram for their catch of elvers, but as the lifestyle of these men is shrouded in secrecy, no one is really sure of the accuracy of this figure.

Should there not be many elvers showing that tide, the elvermen will position their nets in the river, pegged to the bank with two sticks, and then sit back and wait for any stray elvers to swim into and remain in the net. This is called *tealing*. Sometimes, the elvers will drop back with the ebb tide and the nets have to be turned to face upstream. This is known as *sagging*.

Finally, after a long night on the riverbank, the catch is taken to an elver station, where it is weighed and checked before each elverman receives a handsome reward for his hard and sometimes difficult work.

The demand for elvers endures at a time when stocks in the river are dwindling. Not only the Severn, but also other areas of Europe where the elver appears are also seeing greatly reduced numbers. The current fear is that a parasite from Asia is appearing in European rivers and infecting eels.

Fishing for elvers on the bank of the Severn near Stonebench.

An elver fisherman shows how it's done.

A tray used for storing the caught elvers.

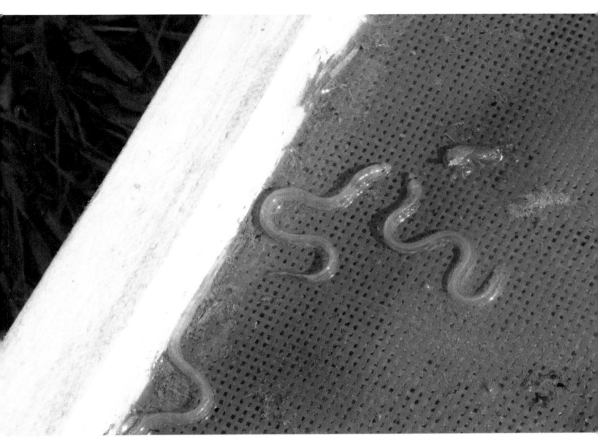

A close-up of elvers in the tray.

Towards the end of the season, the elvers develop small bones, which have the effect of making them look black and unpleasant to eat; this is therefore the time for the elver season to finish for another year.

Should the elver overcome the threat from the fishermen, other obstacles have to be beaten. All along the banks of the Severn there are flood defences with built-in clapper gates that are closed when the river floods. Many elvers get trapped behind the flood defences and one estimate says that 99 per cent of elvers die within the first three months of arriving in the Severn. Please don't feel too sorry for this minute creature as it struggles to beat the elements to get miles upriver into fresh water. They are cannibals. Fill a tank of water full of elvers and years later there will be just one well-fed eel left.

Having reached fresh water in the upper reaches of the Severn, the lucky elvers can move over wet grass and even dig through wet sand to reach ponds and streams to spend eight to twelve years growing into eels.

There are several recipes for cooking a plate of elvers. Here is just one of them:

Ingredients:
    450 g (1 lb) of elvers
    8 rashers of bacon with the rind removed
    2 tbsp of lard
    3 large eggs, beaten
    salt and pepper

- The elvers must be thoroughly cleaned in large bowls of clean water to which cooking salt has been added. Repeat this process several times.
- Leave the elvers to drain and dry in a narrow meshed sieve.
- Melt the lard in a large frying pan, then add the bacon and fry until crisp.
- When cooked, remove the bacon ensuring the fat remains in the pan.
- Add the elvers to the pan and cook until white, making sure they are stirred well.
- Then add the beaten eggs with the salt and pepper and cook until the eggs have just set.
- With the bacon already arranged on a serving dish add the elvers and egg and eat immediately.

Eels

For four or more years, the elver develops into the yellow eel or guelps with an attractive yellow and green colouring on its underside. They remain on the bed of streams and rivers to feed on freshwater food of crayfish, snails, fish eggs and tadpoles. As they mature several years later, they lose their appetite and turn a bluish, almost black colour with a white underside and are known as silver eels or vawsen. The eel has also undergone a physical change as it prepares itself, after developing in fresh water, for the journey in salt water back across the Atlantic. With a small head and small teeth, the body is cylindrical and elongated with a single dorsal fin merging with the tail and anal fin on the underside. For the angler the eels are a nuisance. The angler does not want to catch them and thus spends a lot of the time avoiding them.

As the dark winter nights appear, the eel prepares to face the long journey back down the Severn and out into the Atlantic, before eventually arriving back in the Sargasso Sea – unless they are caught as they pass through Gloucester, which is a favourite place to catch eels. Like elvers, eels command a high price in the marketplace.

It is still a mystery why during the dark nights of autumn between the last and first quarter of the moon, and usually between sunset and midnight, the fully mature eel should migrate down the Severn. When the river has much muddy freshwater flowing down it and the weather is stormy, then there will be an abundance of eels swimming with the current.

In days past, one method of catching the eels was by use of a spear, but that has been forbidden since an Act of Parliament in 1911. The spear was of wrought iron with three to nine flat spines with serrated edges and blunt ends. Mounted on a long ash handle, the spear would be plunged vertically into the mud under the water with the end result being an eel caught between the spines.

Now eels are caught by stretching nets across the river during the period of their brief autumn migration. On average, a single catch can net 69 kilos of eels, but in 1959 a record catch of 1 tonne was recorded. It is doubtful whether this figure will ever be beaten as, like elvers, eels are in decline.

The nets used to catch them are made to suit particular parts of the river. Eel nets are used in conjunction with the Severn punt, and during the autumn these punts can be seen tied to the bank in The Parting. Unfortunately, they are vulnerable to the force of the spring tides and it doesn't take too long to fill them with mud. It only takes a few weeks of neglect before the punt will be buried in all the silt at the bed of the river.

A basket trap is another way of catching eels. Called a putcheon, it is about 1 m (3.3 feet) in length with a mouth 25 centimetres (9.8 inches) wide. The putcheon is baited with a piece of rabbit or lamprey, tied to the bank and weighed down with heavy stones.

Some eels do make it past Gloucester and on down the Severn to the sea. As they move into the sea, their gut dissolves, which makes feeding impossible so they have to rely on their stored energy to get them across the Atlantic. The eel is an incredible creature with an amazing life that started as a dot drifting across the Atlantic and into our shores and rivers.

Here is a simple recipe for how to cook an eel:

Ingredients:
    eel
    cooking oil
    flour
    fresh salad
    salt
    vinegar

- The eel has to be skinned (hard, even for an expert, so please seek help).
- Then it is chopped into 2-centimetre thick steaks and covered in flour.
- Warm the oil in a large frying pan and gently place the eel steaks in the pan and fry for 5 minutes until golden brown.
- Serve with salad and fresh bread.

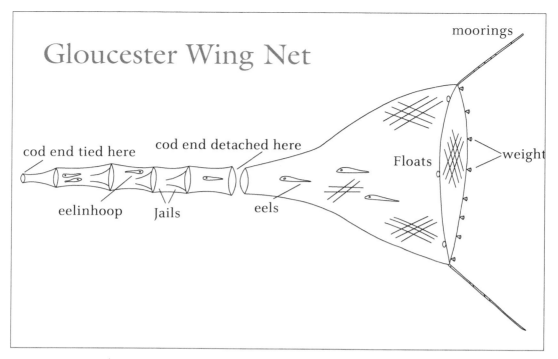

A drawing of a Gloucester wing net.

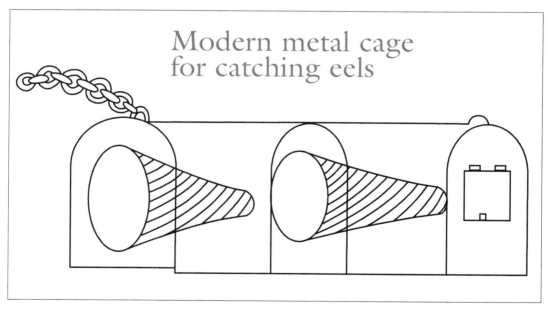

A modern design of a cage for catching eels.

Old type of putcheon for catching eels
Made from willow or hazel

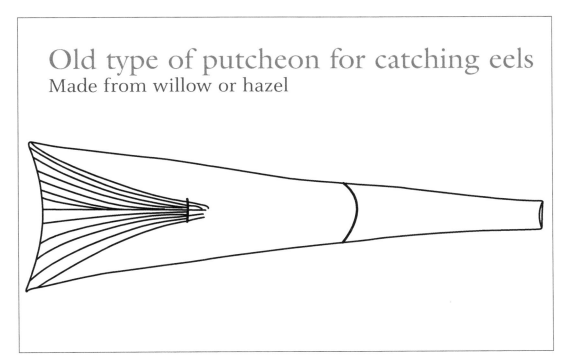

A drawing of an old-type eel putcheon, made from willow or hazel.

Plastic eel putcheon

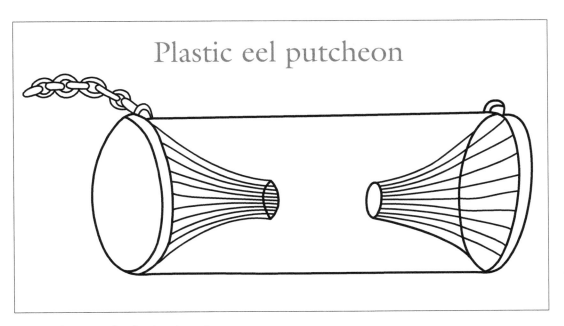

A drawing of a plastic eel putcheon.

Salmon

Like elvers and eels, the salmon is quite a rare species in the Severn today. Until recently, men could be seen on both sides of the estuary standing in the river with their strange-looking nets, patiently waiting for a salmon to appear. Now, only a handful of men operate with the old traditional way of fishing for salmon.

From the sea, the salmon will swim into the Severn Estuary for spawning far up in the shallow and gravelly parts of the river. While in the upper estuary, the salmon will swim close to the surface with their backs out of the water and at risk of grounding on the gently shelving sand of the river. As the tide ebbs out, they are at further risk of becoming stranded on the sand and are at the mercy of the numerous gulls flying overhead in search of food.

To get to their breeding grounds high up the Severn, the salmon have to negotiate the numerous weirs built between Gloucester and Stourport. They will summon up all their energy to swim hard and fast towards the weir, then leap into the air to land back in the water topside of the obstruction. Special salmon chutes have been built on the weirs to assist the salmon in jumping the weirs during periods of low water level.

Salmon eggs when hatched are 5 millimetres (0.2 inches) in diameter and are known as alevins. For the first two years of their life they will remain in the upper reaches of the Severn, during which time they will adopt the silvery hue so characteristic of salmon. They are still only 10 centimetres (4 inches) long and referred to as smolt, but are now ready for the long trip back to the sea.

They remain at sea for several years before returning to the Severn to spawn. They are not called salmon until they are at least four years old. Before this, they are known as grilse, or botchers on the Severn.

The season for catching salmon used to be quite long, from February to August, but now this has been drastically reduced to the period of June to August. Besides men with their lave nets standing out in the estuary, there would be traditional willow baskets, known as putchers, fixed to a strong timber frame built across the main tidal flow of the river. During the close season, the putchers would have to be removed from the frame. Now, with such a short season, the fishermen therefore feel there is not enough time to install and remove them each year.

Hedging (sometimes known as fishing weirs) would be found on the tidal estuary, consisting of stakes placed across the flow of the tide, between which are woven hazel poles (long thin hazel branches are ideal for weaving). The hedging formed a barrier that would guide salmon into the numerous conical-shaped baskets (putchers) that were fitted into the hedging. There could have been as many as 300 putchers inserted in the hedging (mounted in three to four tiers); that number slowly dwindled down to about sixty. The main reason for this was the amount of rubbish found in the river, with plastic bottles and bags continually getting trapped in the baskets.

Each February, salmon fishermen would look for hazel trees to coppice and harvest the poles. Then the men would inspect their hedging and replace any damage caused by the tides with the newly coppiced hazel poles. Meanwhile, they would repair damage to the putchers with willow harvested in March.

During the salmon fishing season, each putcher would have to be inspected twice a day for salmon. As the men waded along the hedging to inspect the putchers, they would have on their shoulder a basket known as a witcher, used to carry their catch of salmon. A more complicated type of basket was also used for fishing, called a putt. It was much larger than a putcher, with three separate sections known as the kype, butt and forewheel. The main use of the putt was to catch salmon, but because of the close weave of the butt section all types of fish were caught, including shrimps.

Other methods of catching salmon have included using stopping boats, putts and drift nets. Until quite recently, stopping boats could be seen at work in the deep pools off Lydney and Gatcombe. They would have been anchored across the flow and fished against the ebb tide. A large net was suspended under the boat and fixed to two large poles, known as rimes. Whenever a fish was felt to strike the net, the rimes in a counter-balance action were raised to retrieve the fish.

With the Severn narrowing at Arlingham and Gloucester, it offered the opportunity to fish for salmon with nets. From a typical Severn punt, one end of a curtain net would be fixed while the other end remained fixed on the bank of the river. As the punt was swept downstream, the net formed into a bag and was then hauled ashore to retrieve the catch. Licences are no longer issued for fishing with these nets (known as long or seine nets).

Now, the only method to have survived into the twenty-first century is the traditional lave net. Used on a falling tide, it demands a lot of skill by the fishermen as they spot a salmon swimming upstream through the shallows. With great agility the fisherman runs through the fast flowing water to intercept the salmon, choosing the right moment to lower his net into the water and then raise it as the fish swims across it.

Keeping alive the traditional way of salmon fishing with lave nets are men from the Black Rock Lave Net Fishermen Association. Black Rock is a picnic site near the village of Portskewett, Monmouthshire, on the shore of the Severn Estuary with commanding views across to the Second Severn Crossing. Bob Leonard has been a lave netsman for fifty-seven years and he knows how dangerous it is out in the estuary. He is quite forceful in explaining that no one should venture out unless they are with someone who knows the river; otherwise, you are putting yourself in an extremely dangerous situation.

Derek Huby, a salmon fisherman, with his folded lave net beside a pair of salmon putchers.

Derek Huby waits patiently in the Severn for a salmon.

A close-up of Derek Huby's lave net.

Old and abandoned salmon fencing near Littleton on Severn.

Fishermen carry their lave nets over their shoulders at Black Rock.

The fishermen head out into the Severn Estuary to catch salmon.

Traditional putchers under construction at Black Rock Heritage Centre.

A completed putcher, ready to catch salmon.

The inside of a putcher.

These men from the Association are not just fishermen; they have also created a small museum in an old fishery building at Black Rock. Visitors are welcome, and brothers Martin and Richard Morgan are only too pleased to show people how a lave net is made and the technique needed to catch salmon. They aim to fish as their forefathers did, using the traditional lave net, which is Y-shaped and consists of two arms called rimes, usually made of willow. This acts as the frame from which the net is loosely hung. The handle is called the rock staff and made of ash or willow. The arms are hinged to the rock staff and kept in position when fishing with a wooden spreader called the headboard.

Salmon are caught during spring tides, as the tide ebbs, using the handheld lave net. The men walk down from the shore with their lave net folded and on their shoulders, as they venture out onto the mud and into the fast-flowing ebbing water, into areas known as The Grandstand, Nesters Rock, or Lighthouse Vear (places you will never see on a map). With water up to their waists, the men stand with their large lave nets open, the ebbing tide rushing through the nets as the men place one hand on the rock staff in readiness to push down with the other hand on the headboard as they wait for a salmon. Having caught a salmon in the lave net, the fish is quickly stunned by hitting it across the head with wooden-type truncheons called a molly-knockers (also referred to as priests or knobbling pins).

Martin Morgan is an expert at knitting the net using a strip of wood and a needle. One of the activities at the museum is seeing him deftly knit the net, and he is only too pleased to let you have a go. The Association are actively promoting Black Rock as a heritage fishery and tourist attraction. Visit www.blackrocklavenets.co.uk to learn more and note the times of when they will be fishing.

The Future

While the world looks for alternative ways of harnessing energy, the Severn Bore is under threat of extinction. Within the last decade, plans have been discussed to build a ten-mile barrage across the Severn Estuary from below Weston-super-Mare to Cardiff.

Such a construction would have a devastating effect both on the bore and on the birds and wildfowl that rely on food from the extensive mud flats in the estuary. There is no doubt that alternative forms of harnessing energy are required, but at what cost? With the Severn having the second highest rise and fall of tide in the world, it is no wonder that the estuary receives so much attention from those looking for a new source of energy. Experts inform us that there is an alternative to a ten-mile barrage.

Nature has provided us with one of Britain's greatest phenomena, which has been enjoyed by many people for many hundreds of years. Hopefully, many more people will be able to enjoy it in the future.

Acknowledgements

I would like to acknowledge and thank the many individuals, who have assisted me in writing this book by providing information. I would especially like to thank my wife, Susan Witts, who not only managed to take some good photographs of the bore at Lower Parting but spent many hours meticulously proofreading my work.

The past few years have not produced the most exciting bores, and looking at the tide table for 2011 I don't predict that we shall see many good ones this year. This makes for much frustration when trying to capture good photographs of the bore. Yet hundreds of people still appear whenever the media announce that a good bore is due on a certain day. I would therefore also like to thank our local press for their interest and coverage given to the Severn Bore.

All the photographs in the book are by the author, except for those on pages 10 (top), 40 (bottom) and 92 (both), which are the work of Susan Witts, those on pages 72, 73 and 74, which are the work of Mark Humpage, and those on page 12, which are published by permission of the Qiantang River Administration of Zhejiang Province.

Further Reading

Arrowsmith's Bristol Channel Tide Table, V. Arrowsmith-Brown, 2011.

Black Rock Heritage Fishery website: www.blackrocklavenets.co.uk

River Severn Tales website: www.severntales.co.uk

Rowbotham, Fred, *The Severn Bore*, David & Charles, 1983.

The Severn Bore and Trent Aegir 2011: A Guide to the Best Locations, Viewing Times and Predictions, Environment Agency, 2011.

Witts, Chris, *The Mighty Severn Bore*, River Severn Publications, 1999.